PRAISE FOR TACTICAL

Having led, been led and seen leadership practised in many ways, Tactical Jazz strikes at what has proven to be critical difference in the quality of leaders and the leadership experience. With Tactical Jazz, while combining the essentials of trust and Occam's Razor, Charles Oliviero provides a sharp introduction to an essential aspect of the art of leadership for those beginning their journey, and also a good refresher for those of us who already have some time in the chair.

Lieutenant-General Christopher Coates
Deputy Commander NORAD

In Tactical Jazz, my friend Chuck Oliviero orchestrates a masterful blend of leadership wisdom, military insight, and improvisational thinking. Drawing from decades of experience in command and strategy, Oliviero challenges conventional leadership models, introducing a dynamic framework that values synchronization, adaptability, and trust. Much like a well-rehearsed jazz ensemble.

With sharp storytelling and real-world examples, Tactical Jazz breaks free from rigid leadership formulas, offering a fresh perspective on how individuals and teams can thrive in uncertain and high-stakes environments. Whether you're in the boardroom, on the battlefield, or leading a creative enterprise, this book delivers a powerful lesson: true leadership isn't about rigid structure. It's about knowing when to follow the score and when to improvise. You can guess what Chuck recommends!

Dr. Darrell Bricker
Global CEO Ipsos Public Affairs

Chuck Oliviero's Tactical Jazz is a relevant and easy read for anyone in the leadership game. His insights mirror many of the ingredients that allowed us to grow from a nascent JTF 2 into an agile, integrated Level 1 CAF formation. We didn't teach troops how to fight as much as we taught them how to think about how to fight. Chuck's central theme of trust —and the tolerance for honest failure needed to create it — is spot on for military and business leaders alike. I remember hearing Chuck teach when I was a young officer attending Army Staff College and his essential truth remains unchanged in that the best decisions are taken by those closest to the problem. The finest leaders set conditions for a diverse ensemble to harmonize!

Lieutenant-General Michael Rouleau
Commander Canadian Special Operations Forces

This book makes a colourful, well-crafted case that, in everything from warfighting to commerce, allowing skilled, well-prepared subordinates to create their own solutions to their problems works better and faster than overly specific, inflexible top-down direction. Rigor without rigidity. Very timely.

Dr. John Scott Cowan
Principal Emeritus Royal Military College of Canada

TACTICAL
JAZZ

TACTICAL JAZZ

by

Charles S. Oliviero

Library and Archives Canada Cataloguing in Publication

Oliviero, Charles S., author

Tactical Jazz / Charles S. Oliviero

Issued in print and electronic formats.

ISBN: 978-1-998501-41-0 (paperback)
ISBN 978-1-998501-50-2 (hardcover)
ISBN: 978-1-998501-42-7 (ebook)

Editor: Phil Halton
Cover Design: Pablo Javier Herrera
Interior Design: Winston S. Prescott

Double Dagger Books Ltd.
Toronto, Ontario, Canada
www.doubledagger.ca

I dedicate this book to the men and women I have known through my attendance at The Royal Military College of Canada. Through their examples, I have come to understand the true meaning of our College motto:

Truth Duty Valour

Jazz is about freedom within discipline.

Dave Brubeck | American Jazz Musician

We look at the way things tend to go and leave the individual more freedom to use initiative.

Prince Frederick Charles of Prussia

LEAVE THE GUN; TAKE THE 'A' TRAIN

AFTER COLONEL SKIP HAD POURED THE DRINKS, he leaned on the bar with both elbows, appearing lost in thought. He stared out over the lounge where the majority of his troops were sitting and relaxing in twos or threes.

"Colonel, you look pensive."

"Pappy, I was contemplating how well everyone has coalesced into a unit in such a short time. You men have done great work."

"Aye, the squads are runnin' like them Swiss watches you like. That they are."

Skip took his elbows off the bar. "God, I hope not."

Pappy was startled at the comment and Harry attempted to intervene. "Now you've done it, Pappy. We're going to spend the rest of the evening discussing jazz."

Pappy, not to mention everyone else, was puzzled. "I don't know what the hell you're talkin' about, Harry."

"Gentlemen, please grab your drinks," said Skip. "We'll reconvene in my office in five."

1

Harry grabbed his glass and looked at the bewildered men with a grin as Skip headed back to his office. "What did I tell you?" he said and followed Skip.

As the men settled back into their chairs, Skip booted up his computer. He began to play a song at very low volume. It was a live jazz performance. Jules began to speak, but the Colonel raised his hand to quiet him as he slowly raised the volume. Everyone sat and listened for several more minutes, after which Skip lowered the volume, letting the music play on. "Question or comments?"

Pappy had begun this whole process, and he wasn't about to put his foot in it twice in a row. Jules finally spoke. "Mon Colonel. I am not a big jazz guy. I recognize the song, I think. Something about taking the train, maybe, but I do not see why we are listening."

"It's actually called 'Take the "A" Train,' by the late Duke Ellington," said Skip, "although this is not his band. This is a recording of the Charles Mingus Sextet. Mingus is on bass, with Eric Dolphy on alto sax. Johnny Coles is on trumpet. Dannie Richmond is on drums and Cliff Jordan on tenor sax. Obviously, you can't see any of the musicians, but if you could, you would see that none of them are reading from musical scores. They are all sitting on a stage and jamming with each other. If you listen, you can hear each of the artists come in and go out of the song almost at random. Did anyone notice that at one point everybody stopped playing to let the pianist, Jaki Byard, play solo? Then the drummer stepped in, and soon all six musicians were jamming again." Skip surveyed the group. "This, gentlemen, is the opposite of a Swiss watch. It's how I expect you to operate."

Alex and Harry were notably silent. They had been down this road with Skip before, and they each decided to let the others commence their own voyages of discovery. Breaker spoke up. "Okay, Colonel. So what? Those names mean nothing to me, and I'm not sure exactly where you're going with all this, to be honest. Sorry."

"No, don't apologize, Breaker. There's nothing to be sorry about. If you aren't into jazz, then chances are you've never heard of Chuck Mingus, and you don't know that he was one of the greatest jazz musicians of the twentieth century. But this is not about Mingus, or his band. It's about unit cohesion and operations."

"Unit cohesion?" said Breaker.

Skip looked at the men and thought for a few seconds. "Let me explain."

The real voyage of discovery consists not in seeking new lands but in seeing with new eyes.

Marcel Proust | Novelist and Critic

THE CHALLENGE

THE SHELVES OF BOOKSTORE self-help sections are filled with tomes about leadership. A quick search on Amazon Books turned up an incredible 60,000 titles! What does such a quick exploration tell us? A great deal. Clearly, there is an ever-present demand for such literature. Further, there is a broad spectrum of thought and theory on what constitutes good leadership. And this is good.

Evidently, there is no single best practice for a leader. Leadership (and management) are deeply human activities, and as such they contain elements of both art and science. Every leader is an individual who must combine psychology, sociology, language skills, empathy, sympathy and a dozen other arts and sciences on a daily basis. Such a task is no mean feat under mundane circumstances, and the challenge increases in complexity exponentially as stress and importance grow. The good leader knows that he, or she, cannot take as given any single mode or method. The good leader appreciates that the mission of a leader is dynamic. That it changes with every situation and with every group that must be given direction and guidance. In some cases, one might adopt the patient, non-violent methods espoused by the Dalai Lama. In other cases, the rough and ready approach of George S. Patton Jr.

One core truth abides: good leadership is the critical component of all successful operations, and it makes no difference whether the organization is a military unit, a government department, a hospital, a classroom, a bank, or a retail hardware store. But this book is purposely not like any of those 60,000 volumes already on the shelf; this one is not about rules and tips. It's about a concept. It's about a simple concept that can be the seed that grows into a magnificent forest. This concept is synchronization, but not just synchronization. It is about much, much more.

The scenario of the jazz sextet above comes from my novel The Cohort, but it's actually a manifestation of a key point I made in one of my earlier books (Auftragstaktik: The Birth of Enlightened Leadership). The point I was making was that there is a widely-held — and thoroughly mistaken — belief that military leadership is rigid. That it is formulaic and dogmatic. When I left the Regular Army after three decades of service, I was repeatedly bemused by the reactions of civilian leaders and managers who were surprised to discover that ex-military leaders like me were rational, open-minded and tolerant of subordinates who directly questioned orders and direction. I frequently heard of managers who were seriously annoyed by subordinates who questioned them. These people were invariably uncomfortable with strong subordinates, saying that any questioning or pushback from subordinates made them uncomfortable. In my mind, they were not good leaders, and in my experience, strong subordinates always made strong teams.

The concept of synchronization is not new, and I have taught and spoken about it for decades, but what has prompted me to write this short treatise was a comment by an old friend. He was never in the military and sent me a note because he was intrigued by the passage from The Cohort. The novel's protagonist is a retired armoured cavalry colonel, who is leading a special operations unit, and he's explaining his concept of operations for an upcoming mission to his men. He's attempting to get them to step out of their comfort zones and to think of operations in a different way. That's his challenge (and yours).

A word of caution: this short exposition is not intended to be an in-depth investigation into the subject, for that would require much more study. Instead, it is an introduction to an idea that may initially seem counterintuitive. Nonetheless, if you are a leader or a manager in any industry or profession, I hope that you will find this both interesting and useful.

You 'musicians' of Mars must not wait for the band leader to signal to you. You must, each of your own volition, see to it that you come into this concert at the proper time and at the proper place.

George S. Patton, Jr. | American Army General

MUSIC OF MARS

COLONEL SKIP NEEDED AN ANALOGY that was more easily relatable for his men. "Listen," he said. "There are two ways to look at a military operation. The first, and more conventional way, is that everyone in the operation has a part to play, and they each do exactly what they're told. It's a giant machine. The Soviets took this idea to ludicrous extremes, not allowing any initiative or discretion below divisional levels. Conversely, General George Patton described combat in terms of a symphony. He called it the music of Mars, where each weapon system played its part of the musical score and the commander, the general, conducted it, so to speak. Decades after Patton described it this way, the US Army dubbed his idea synchronization. Further, they developed a tool for commanders and staffs called a synchronization matrix so all the combat power could be coordinated to create this synchronization. Roger so far?"

Everyone nodded. Harry and Alex shared a knowing look.

"The second way to look at an operation is that it's like playing in a jazz combo. Each musician knows his own instrument, and he doesn't need the conductor to tell him when or how to play. He listens to the music and feels the rhythm, the tempo, the emotion, and joins in when, where, and how he sees fit. There are several key elements to appreciate here. Each player needs to understand the whole piece, but he must also subordinate his own

playing to the goal of building synergy, to the beauty of the melody. See the difference?" Skip looked at his Orders Group.

"Sorry, Colonel, but no, I don't," said Breaker.

"Breaker, there's still no need to apologize." Skip paused for a moment. "Let me try it another way. In the first methodology, the synchronization is planned or even forced: it's a program: it's a process. In the second methodology, it isn't a process. It's an outcome." Skip paused to let the idea sink in. "If the analogy of jazz doesn't do it for you, then think of hockey or any other team sport. We've all played on or watched a sports team that was in the moment. Athletes call it being in the zone. Every pass works. Every shot scores. In music, this happens when every player submits to the tune. Every individual is in synch with every other player. It's where the whole is greater than the sum of its parts. It's rare, but it happens." Heads were now nodding.

"Okay," said Breaker. "I think I'm beginning to get it. So, can you please explain to me why that isn't like a beautiful Swiss watch? I mean, everything's working perfectly, right?"

Skip smiled momentarily. "Because a finely tuned watch can only function if every aspect of its mechanism works exactly as predicted. That's fine for Swiss watches. They live in a closed environment. But that's practically impossible for combat operations. Absolutely nothing in a military operation is completely predictable." He looked at his men. "You're all highly experienced soldiers. You know what I'm saying. To return to the watch analogy, a single grain of sand will stop a ten-thousand-dollar Rolex, and in our case, we're trying to run those watches in a sandstorm with no backs on them."

Skip could see that his message was sinking in. He was impressed that all the men assembled here had the mental agility to reconsider decades of military indoctrination. "Listen, it wasn't my intention to lecture you all on the Zen of Operations," said Skip, shutting the laptop. "Let's call it a night, everyone." He looked at his two subordinates. "Alex, Harry, please stay behind."

Alex and Harry sat while the rest of the men filed out of the office, and Skip closed the door. "Sorry, guys," he said with a wry smile. "I guess the nutty professor escaped from his straitjacket, and I lost control of him again. I hope I haven't messed with their heads. That was certainly not my intent." Skip looked concerned.

"Don't sweat it, Colonel," said Alex. "They get it, believe me. I'm sure Harry agrees with me. The men are much closer to being a jazz sextet than they are to a parade square formation, trust me." Alex had a broad grin on his face. "Even if they never heard of Chuck Mingus."

"Alex is right," said Harry. "Besides, the nutty professor needs to escape now and again, and I, for one, always learn something new when he does." He too was grinning.

Skip was shaking his head. "If you say so Harry," he said. "I worry sometimes that my desire to stretch their minds and force them to look at things with a new perspective might be counterproductive."

"Not a chance Skip," said Harry as Alex nodded his agreement.

15

The creation of a thousand forests is in one acorn.

Ralph Waldo Emerson | American Philosopher

THE ACORN

WHETHER IN INDUSTRY, COMMERCE, government, medicine, or the military, every successful new endeavour begins with an idea; with an initial distinctive spark that ignites a bit of intellectual kindling in someone's brain. Once that fire is lit, sometimes it spreads to another part of your brain — and if you are perceptive, or maybe just lucky — that small cerebral fire begins to grow and spread.

That process is what I hope to initiate and then foster. This book is intended to be that initial distinctive spark, and it's my sincere hope that it ignites those many small pieces of my readers' brains. Why do I want that? Because it is my conviction that it is a fundamental first step in achieving success.

One of the many "chicken or egg" arguments is whether technology drives progress or whether progress drives technology. I fall firmly down on the side of progress being the driver — but only insofar as we understand progress to be the encouragement and adoption of new ideas.

This initial spark then becomes a small but potent initiator. A nascent idea that serves both as a forcing function as well as the foundation for new strategy or action. Very much like a single acorn that has the potential to become a stately oak forest, this notion underlines the importance of the intellectual tools of imagination, insight, and conceptual innovation as precursors to effectiveness. Developing these tools (some may consider them skills) is important for all of us, but particularly critical in leaders at all levels.

A winning strategy almost never materializes fully formed. Instead, it evolves from a simple yet powerful idea. This idea might stem from a keen observation, a new perspective on a problem, or a spark of inspiration. It then acts as the nucleus around which a strategy is built, providing direction and coherence. But it doesn't stop there. Both the creation of the idea as well as the development of that idea into a strategy needs to be nurtured, cultivated and developed.

For instance, in sports, a coach might notice an opponent's defensive weakness and develop a new offensive play to exploit that weakness. In business, a groundbreaking product concept might arise from identifying an unmet customer need. In medicine, a new technology may offer the possibility of altering treatments to combat disease and illness. In all three cases, the initial idea — the spark — initiates and then guides subsequent planning and action.

This principle reminds us that success starts with the courage to conceive of something original, the discipline to develop it, as well as dedication to nurture it into a concrete plan. By valuing and cultivating these small but transformative ideas, leaders, individuals and teams lay the groundwork for impactful, winning strategies, tactics and processes.

In this book I will introduce some concepts, which were embraced by soldiers and thence migrated into military theory. But that should not deter readers who are non-military, for it is merely the point of departure. These concepts are neither uniquely suited to nor restricted to military operations. They are intellectual constructs and therefore applicable to any and all institutions, businesses and organizations that seek to improve and grow, whatever they may be. All can greatly benefit from the embrace of these ideas.

Wherever you work, if you are a leader — at whatever level — this book can be your personal point of departure to achieving meaningful outcomes, for you personally, for your team, and for your organization.

I've always been interested in science. I used to take watches apart and clocks apart, and there's little screws, and a little this and that, and I found out if I dropped one of them, that thing ain't gonna work.

Herbie Hancock

SYNCHRONIZATION

IN THE POST-VIETNAM WAR ERA, the US Army went through an intellectual renaissance. By the early 1980s, their *Active Defense* doctrine was replaced by *AirLand Battle* as the army's conceptual warfighting framework. *AirLand Battle* emphasized close coordination between ground forces acting as an aggressively manoeuvring defence, and air forces attacking enemy rear-echelon units close behind the in-contact enemy forces.

A critical aspect of this new doctrine was that all forces be synchronized. The US military seized upon this idea wholeheartedly. As originally defined, synchronization was the ability to focus resources and actions in both time and space in order to maximize the creation of combat power at any given point, thus creating synergy. Although the definition did not change fundamentally, how it came to be applied soon bifurcated between the the Army approach and the US Marine Corps approach. The two organizations did not agree on how this concept of synchronization was to be achieved. The former side saw synchronization as a *process* while the latter saw it as an *outcome*.

AirLand Battle encouraged all commanders to build synergy with their subordinates, peers, and superiors through the leadership philosophy of *Auftragstaktik,* which the Americans awkwardly translated into English as *Mission Command.* The USMC applied this concept by training all leaders to think independently within a coordinated whole. The US Army took a slightly different approach. They preferred the use of a command and staff tool called a "synchronization matrix."

Obviously, the fact that two large, successful fighting organizations saw it so differently would imply that the concept of synchronization could be achieved in more than one way. At the lower tactical levels, I remain strongly inclined to favour the USMC interpretation, that it is an *outcome.* But I can certainly appreciate that at higher levels it may indeed be a *process.*

Let's consider some commonalities between these two views. In both cases, commanders are key. Whether forcing synchronization through a matrix or achieving it through a well-coordinated leadership team, both views see synchronization as a command function. The problem for commanders and their staffs — this would particularly apply in business — is that at the lowest levels, attempting to use command synchronization tools like matrices becomes an incumbrance. In other words, synchronization works best when it occurs spontaneously, and organically. Large organizations, working at a strategic level, can induce synchronization through a synchronization matrix or some equivalent computer tool, but at low levels and in small organizations, forcing the use of such tools can often become a burden and thereby be counterproductive.

In my own experience, for the low-level leader, synchronization can best be understood by using the analogy of jazz. Think back to Skip introducing this idea by making his subordinates listen to the jazz sextet. The special operations unit led by Skip is composed of fewer than two dozen men, so Skip wants his subordinates to appreciate that synchronization must occur spontaneously, freely.

Let's dig deeper into the analogy. A jazz ensemble may have a written score to get them started, but any lover of jazz will tell you that it doesn't get good until the musicians wander off the pre-ordained path. The musicians have to find their "groove" and once they do, then the group coalesces. They get "tight", and the music takes on an ethereal quality that cannot be gained from methodically following a written score. If you aren't a jazz lover, then look at sports. Whether you are a high-performance athlete or just someone who plays an occasional pick-up game of basketball or ball hockey, chances are that at least once in your life you've felt the sensation of being "in the zone." Athletes use this term to describe the near out-of-body experience when everything they do, every play they make, goes perfectly. The pass to a teammate is perfect, the kick on net hits exactly where you wanted it to, the throw to first base is perfectly timed for the double play that wins the game.

How do you get to the sweet spot or be in the zone? In his book *Outliers,* Malcolm Gladwell established what he called the 10,000-hour rule. The rule says that only through constant practice can you achieve professional proficiency, and he reckons that it takes more or less 10,000 hrs. Psychologist Anders Ericsson has pretty firmly debunked Gladwell's "rule" as being a misinterpretation of the data, but it still has value. It is less about the precise number of hours than about gaining proficiency through long-term, dedicated study and practice. What this means to you is that you must constantly be training yourself and your subordinates to work as a team. They need to be able to get inside your, and each other's, decision cycles in order to understand what you as the leader expect – even when you are not there.

Of course, combat is not a game, nor is it a jazz set on a stage. With adrenaline flowing and time compression weighing heavily on you and your subordinates, every decision by every member of your unit can become a life-or-death moment. The only way to prepare your troops for these moments is good training. Remember Erwin Rommel's adage about training being the best form of troop welfare. If you train them well, and if you lead them well, you'll achieve synchronization.

Business likes to use the vocabulary of combat, but it is rarely a life-or-death struggle. Nonetheless, in a multinational cutthroat commercial environment, not being as efficient or as productive as possible could lead to the loss of your livelihood or even the demise of the entire corporate enterprise. Consider Enron's bankruptcy in 2001 or Bear Stearn's failure in 2008. So, whether you are a military leader, an entrepreneur who is building a business, or a corporate executive who wants to see your executive team perform at a higher level, synchronization may well be the key that unlocks the door to achieving your goals.

Let's assume that it is. So what? How do you begin?

Start with yourself. You cannot lead your team into a new venture if you do not know where you are going. You need to educate yourself in the basics. Then, you can decide how you will bring along your subordinates. In the meanwhile, be in moment as opposed to looking for what scientists call computational reducibility predications. In other words, do not try to create a formula that allows you to plug in what you know in order to get an outcome. Humans are not like that. Internalize the process. Understand that every situation is unique, even if all of the parameters appear to remain unchanged. Each time you gather your team to advance another small step, everything may look the same, but you may not get the same, or expected, outcome. This is the real reason that I personally do not like the synchronization matrix. It is a scientific solution to what is essentially a human issue.

Let's consider one way to put this concept into practice: the German military notion of *Auftragstaktik*.

You make different colors by combining those colors that already exist.

Herbie Hancock | American Jazz Musician

ALL OR NONE: AUFTRAGSTAKTIK

AUFTRAGSTAKTIK IS A PHILOSOPHY of military command emphasizing decentralized decision-making and initiative within the framework of a commander's intent. Originating in the Prusso-German military, it encourages subordinates to internalize the overarching goals of a mission rather than follow rigid, detailed instructions. Commanders provide clear objectives and the desired end state, leaving the *how* to achieve the goals to the discretion of lower-level leaders. This approach fosters flexibility, adaptability, and rapid decision-making, especially in a complex and unpredictable situation. By trusting subordinates to act independently but within the leader's intent, *Auftragstaktik* promotes efficiency, innovation, and unity of purpose in achieving objectives. This book is not about *Auftragstaktik,* but we cannot discuss synchronization without appreciating that it is native within this leadership philosophy. If you want to know more, look at another of my books: *Auftragstaktik: The Birth of Enlightened Leadership.*

Let's begin with the star-shaped model. The five points of the star represent the major tools of *Auftragstaktik*. The centre of the star is the fundamental base that *must* exist if the tools are to be used effectively. But the model isn't as simple as it seems. The tools can be used independently or in combination. Further, although trust is an

AUFTRAGSTAKTIK

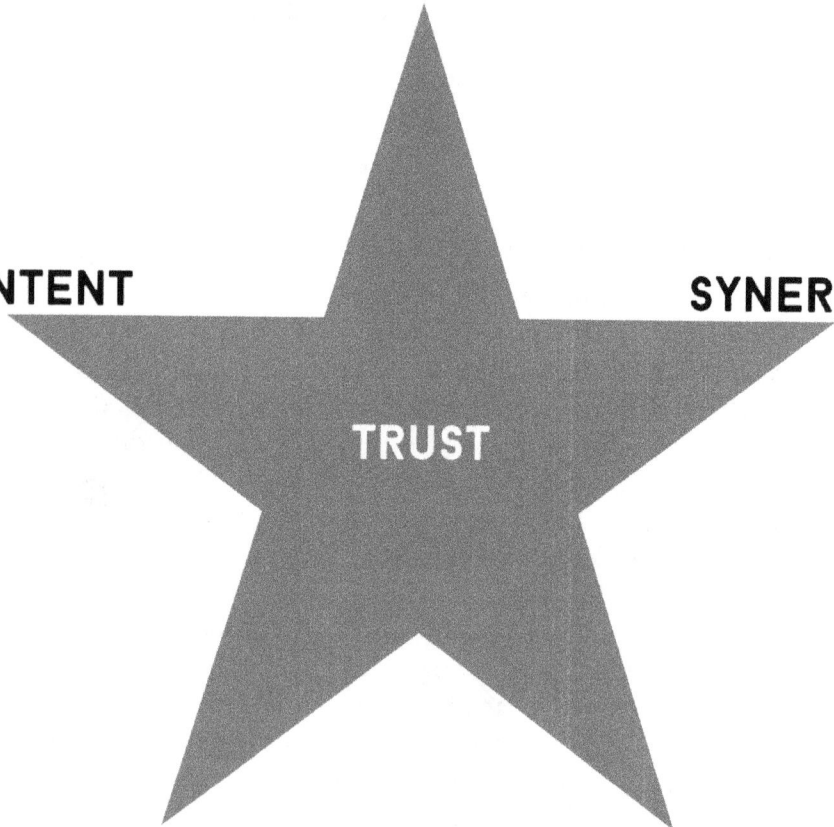

critical requirement, the tools can be used in the *absence* of trust in order to *build* trust.

Synchronization

Defined as the activity of two or more things at the same time or rate, synchronization is the basis that facilitates synergy. Without synchronization, efforts are disjointed, failing to achieve the desired emergent, amplified effects that define synergy. Together, these concepts highlight the power of aligned and cooperative systems to achieve extraordinary results.

Synergy

This is the interaction or cooperation of two or more organizations to produce a combined effect *greater than the sum of their separate effects.* The final part is crucial. It isn't that one and one make two; it is that in some cases, one and one make *three* or even *more.* **Synchronization** and **synergy** are obviously closely interdepended concepts, both rooted in the idea of interconnectedness and cooperation. But they operate in slightly different ways and contexts. Together, they describe how coordinated processes or systems can amplify effectiveness and create outcomes greater than individual contributions.

Initiative

At its simplest, it's the ability and willingness to act without waiting for explicit instructions. It involves taking responsibility for pursuing tasks, suggesting improvements, or addressing issues to advance organizational goals. Initiative is characterized by self-motivation, resourcefulness, and the capacity to think and act independently while aligning with the organization's objectives. It often leads to innovation and efficiency. Leaders play a key role in fostering initiative by empowering employees, providing support, and recognizing proactive efforts.

Cohesion

This is the degree of unity and collaboration among members of a group, enabling them to work effectively toward shared goals. Cohesion reflects the strength of interpersonal bonds and the alignment of individual objectives with collective objectives. High cohesion fosters open communication, commitment, and a supportive environment where members feel valued and are motivated.

Intent

The leader's intent is the desired purpose of a mission, project, or task that provides the guidance and the end-state to be achieved. Within *Auftragstaktik*, intent empowers independent decision-making. By communicating the *why* of a mission, rather than prescribing every detail of the *how*, the leader's intent fosters flexibility, creativity, and individual ownership among team members. It ensures alignment of purpose and unity of effort all the while fostering adaptability in dynamic or complex environments. The effective use of the leader's intent fosters trust, accountability, and resilience, enabling teams and individuals to navigate uncertainty and deliver outcomes aligned with organizational goals.

Trust

Trust lies at the heart of *Auftragtaktik's* efficacy. In its simplest form, *Auftragstaktik* refers to the mutual trust between superiors and subordinates, where a superior sets goals and gives subordinates free rein to achieve them. In its fullest and most elevated applications, it makes all members of an organization participatory stakeholders in the achievement of a mission. The concept is based upon the pillars of the subordination of self to a superior's intent, independent action, and freedom of action at all levels by all leaders. Put simply, *Auftragstaktik* can be summarized as trust, training, and simplicity.

Summary

Obviously, the star model is highly simplified, and if you understand models, then you will know the maxim that *all models are wrong, but some are useful.* That is the case here. The star gives us a snapshot of how all of the aspects of *Auftragstaktik* relate to each other, while keeping in mind that in this short investigation we are specifically interested in synchronization and how it relates to synergy. Naturally, we cannot investigate the tools in isolation, but let's review with a focus on synchronization.

- Synchronization enables synergy. For synergy to occur, the components of a system must be synchronized in time or action

- Synchronization aligns components to cooperate effectively, thereby setting the stage for synergistic interactions

- Synchronization minimizes friction, allowing systems to focus energy on synergistic outcomes

- Synchronization promotes overall efficiency, since aligned deadlines and tasks prevent overlap and waste.

Synergy is then the *manifestation of* interconnectedness in your organization, and synchronization is the *internal experience* of that interconnectedness. Together, they suggest an organization, whether military, civil, governmental, or commercial, in which collaboration and alignment lead to profound outcomes.

In theory there is no difference between theory and practice. In practice there is.

Yogi Berra

THEORY
MEETS
REALITY

TO EXPLOIT SYNCHRONIZATION'S manifold advantages, you need to embrace *Auftragstaktik*. Where to begin? To borrow from Ernest Hemingway's 1926 novel *The Sun Also Rises*, "Gradually, then suddenly." Consider the star model. All the tools interact with each other in an environment of trust. Conversely, to build trust, the tools can be used to grow trust. Your subordinates need to know that they will not be punished for using their initiative. Empowerment is the key, but it takes work. First, you need to make them understand that mission success is shared, and that all members must be invested in it. You also need to demonstrate it. You must show them that *your* mission is also *their* mission. If *you* succeed, it is because *they* were invested, and in order to promote that investment, they must not feel afraid of making an error. Honest errors must never be punished, especially if they were made with the purpose of achieving the mission in the absence of a leader or of clear direction. They need to know that you trust them after which they will trust you.

But theory and reality don't always align. It is one thing to explain how something *might* work or how it *should* work, but can it *actually* work in a concrete setting?

We've all seen those rare occasions when everything just clicked. Let me now share some personal examples. Keep in mind that these events were not accidental. Each was the result of having a highly trained team committed to excellence and willing to "play without a score."

Tactical Jazz

Because I grew up in a single army regiment, by the time I was given the privilege of command, most of the officers and non-commissioned officers knew me and my leadership approach. Shortly after taking the reins, I deployed the regiment on the annual NATO training exercises. After some excellent refresher training conducted by my subordinate commanders, it was time for my own exercise, a battle group advance to contact. This involved my entire force of almost 1,000 soldiers. The enemy was a sub-unit of our regiment, and they were keen to show us up. Once the battle group shook out, we started advancing.

Within minutes, one of the 19-tank subunits (B Squadron) made contact. Normally, a battle group attack – even a quick one – took upwards of 90 minutes. Using the drills that I had professed for years, everyone swung into action without orders. I radioed B Squadron to tell them where I wanted the fire base established. The commander radioed back: "I thought you would want us there and will be ready in two minutes." In other words, he had not waited for orders. He started moving in anticipation and felt he knew where I would want him. I then told one of my other 19-tank subunits (A Squadron) where I wanted to gather everyone to launch the attack and that he should begin his move there. His reply: "I have already secured it and am awaiting your arrival." I then grabbed my last subunit (C Squadron), which had been waiting because the squadron commander knew his job would be to follow me into the attack, and we were off. Naturally, this was a peacetime exercise, and the enemy was not firing back with real bullets, but we were on the objective in less than 45 minutes. It was one of the proudest moments of my regimental life. With almost no orders, my subordinate commanders and their subordinates had all acted on their own initiative. By the way, all three of those officers went on to command armoured regiments and two of them retired as major generals.

Follower Empowerment

On the first day that I took over as the Chief of Staff of the Canadian Army Command and Staff College, the army's profession school for officers, I gathered all of the college support staff in the main lecture theatre. From gardeners to cleaning staff, from military officers to secretaries and librarians, I gave them a ten-minute simplified lecture on *Auftragstaktik*. Then I announced "I hereby empower each and every one of you to decide on the spot if a decision is required. If your decision turns out wrong, you will not be punished or chastised. However, if you saw the need for a decision and did not decide to make the situation better, I will be disappointed. Questions?"

There was silence, then one of the civilian secretaries in the production office asked how that affected her. I offered that perhaps she might be working on a course package that needed to be mailed out to future students and that someone from the printing office had called needing urgent permission to print extra copies to mail out that day. The request was time critical, but those course packages were the strict purview of the Operations Staff. It was Friday afternoon, and they were gone. You could simply tell the printing office to wait until Monday. It would cause a delay, but it wouldn't be your fault. Better still, knowing that the need was both legitimate and urgent, you could give the printing office permission and then put a note on the Operations Officer's desk explaining your decision. She beamed.

"Really?" she said.

"Really," I replied.

For the next few weeks as I walked around the college talking to the staff, I was constantly engaged by members who were thrilled to have been empowered as never before. It was infrequent that any of them had to take decisions as I had laid out, but occasionally it happened, and I always made a point of praising them publicly. A couple of times poor choices were made but again, I praised them for their initiative, not for the outcome. It didn't take long for the staff to understand that they really did have the power to make the college work more efficiently and most embraced that power.

There is no reason that you and your subordinates, whatever your endeavour, cannot train and be trained in this way. Your people are capable, intelligent and most importantly, they have great stores of initiative. To foster that initiative, you need to demonstrate your commitment to *their* success. You need to exhibit daily that you are willing to accept more independence from your subordinates, and that tasks and goals are *shared* responsibilities; that you expect them to carry out those tasks and goals even in the face breakdowns and lapses in communications or the removal of any leader in the decision chain. But remember the model. First, you need to earn their trust.

Building Trust

When I was a young lieutenant, I was deployed with my regiment on a UN peacekeeping mission as a troop leader. I had five senior non-commissioned officers (NCOs) and over thirty-five soldiers. About three months into the tour, I was due for a long weekend off. There had been an incident on the border that the Commanding Officer (CO) had specifically warned me to watch closely. Anxious to go on leave, I gave it only a cursory investigation and worse still, without analysis, I simply passed his instructions to my deputy, a senior warrant officer. Then I went to the leave center. Upon my return, I was told by my warrant officer that the situation had blown up while I was away and that the CO wanted to see me. Immediately. The CO was most upset; worse, he was disappointed. I had let *him* down, and he wanted to know why my troops had let *me* down. Looking back, I now know that this was a test. He was testing me to see if I would blame my subordinates. I didn't. I accepted responsibility for the failure and praised my men for doing the best they could, given that I had left them poor instructions. He looked at me for a moment, then he dismissed me.

I returned to my troop. My warrant officer and NCOs were waiting for me. They asked if I had been fired. No. Extra Orderly Officer duties? No. I told them that I had taken full responsibility and that the CO had accepted my apology. Something changed that day in my troop. The NCOs, who had previously been respectful but cool became more respectful and more loyal. I had taken it on the chin, and they respected it.

The CO was indeed disappointed in me, but he also knew that he had given me some of the best NCOs in the regiment and that they could have done better. I never blamed them because it was *my* job to ensure *their* success, and not the other way around.

*So I let go of the reins and told them,
'Run! Do it your own way!'*

Psalm 81

LETTING GO OF THE REINS

AFTER LEAVING THE REGULAR ARMY, I managed a large civilian training organization. Through a friend, I managed to get a foot in the door of a world leader in energy production. They wanted some emergency response training for their on-site supervisors and managers. They had invested millions of dollars in new infrastructure and equipment, but wondered if old mindsets would work with all of this new technology.

I suggested a week of small tabletop exercises (TTXs) with a hand-picked team of experts to train the target audience in new leadership and management mindsets. I knew exactly how *I* would conduct the TTXs, but considering the size of the training audience, it would not be practical for me to do it all. The VP Operations accepted my proposal.

My deputy and I selected a group of trainers we could trust to conduct small TTXs within a larger exercise scenario. I chose a dozen names and together he and I whittled the list down to a half dozen. Then I asked him to create the overarching scenario. He ran it by me for approval. I didn't change anything.

I hired the selected trainers (each was an independent contractor).

I called them together and explained how important this series of TTXs would be. I held out the hope that the client would be pleased and hire us to create a larger long-term training plan for a large-scale event. I laid out my philosophy, the goals, the training objectives, the methodology and the training scenario. I then explained the concept of operations for the week we would spend with the client at their site as well as key timings and activities (daily gatherings, after action reviews, both by syndicate and in plenary, group discussions etc.).

In the hotel near the facility where we were staying, I gathered the instructors one last time. I told them I knew they were all expert and experienced, and that they had free rein to conduct the TTXs as they saw fit, within my overall concept. I told them I was confident that they would impress the client. Then we went to the site where we met the training audience. I gave a plenary briefing where I introduced each of the syndicate facilitators. They broke into their syndicates and began. I kept away from the syndicates except to visit for a few minutes every morning and afternoon.

Each of the facilitators had his own methodology, with no two TTXs being identical. It didn't matter. All the facilitators worked toward the goals I had set. We gathered at the end of each day to share ideas. They asked for feedback from me, and I always said the same thing: "Keep doing what you're doing. It's working."

At the end of the training week, the VP Operations took me aside and told me that he wanted to hire my organization to create a year-long training package for him, and to then conduct the training. The contract was huge.

I hosted a celebratory dinner that night with my small team and praised them for "hitting it out of the park." I asked them if they would like to stay on to fill the contract we had just won. Everyone signed on.

Check Your Ego at the Door

Before winning the contract above, an old army buddy, who had had been seconded to government, suggested that my organization should be engaged to help train everyone involved in hosting the 2010 Vancouver Olympics. It was a massive undertaking with the need to create a training plan that coordinated almost 130 diverse government, police, security, medical, sports and military agencies, including synchronizing our military with the police and the US military. Luckily, we started almost two years before the event.

The number of events that needed to be conceived, designed, produced, and conducted was exceeded only by the number of meetings and conferences that were called by the national organizing committee, which reported to the Prime Minister's Office. Early on, I was asked to join that committee in an advisory capacity.

Once the training plan was accepted by the national committee, we commenced training. With national pride, political sensitivities and international scrutiny ever-present, I was extra vigilant to ensure any missteps were few and minor. I had selected the Activity Leads for their experience, but also for their strong leadership and I trusted them completely.

Several months into the training, we held the main planning conference for one of the first big exercises. There were participants from across Canada, the US and observers from Britain (the next Olympic host). At one point, there was some minor confusion, and a member of the national committee asked me for a decision. I called over the Activity Lead for the training event and asked him to clarify. The committee member was puzzled. "Aren't you the boss?" he asked. I explained that I was responsible for the entire Olympic training series but that this particular exercise, like every exercise, had an Activity Lead. It was his show and in fact, for this event, I was just a bit player in the exercise control organization and little more than a duty officer. That completely confused him. How could I be in charge of everything but subjugate myself, several levels down, to one of my own subordinates?

"It isn't that complicated," I said. "Of course, I'm responsible to the committee to ensure success. But in this training event, I am merely a controller responsible to the Activity Lead."

"But he works for *you*," said the committee member.

"That's true; and *I* work for him," I said.

"Why would you do that?" he asked.

That was when I explained that although I maintained an overwatch and would not allow any event to go off the rails, the Activity Lead was in charge of this particular event with the freedom of action to ensure success. The reason I had chosen him was simple: He was better at this type of training than I was. I just happened to be ultimately responsible. The committee member thanked me, but I could see that he was a bit worried, and not convinced.

Postscript

A few days later, the committee member came back to me. As he had watched the training progress, he better understood and appreciated my methodology. He had heard of this this kind of leader-follower-situation, but had never before seen it in practice.

As well as a myriad of small training events, there were three major non-military training exercises to prepare for the Vancouver Games, and they were three of the largest security exercises ever conducted in Canada. Each of the three Activity Leads performed superbly and they made history. Each of them received a recognition award from Canada's Clerk of the Privy Council, an honour never before bestowed upon someone who was not a civil servant.

Nota Bene

I cannot move on without a note of warning. Everything I have said above presupposes that the individuals with whom you are working are both professionally oriented and skilled at what they do. It is not possible to build an organization of unskilled amateurs and then immediately give everyone the freedom to decide on their own what to do.

It should be obvious that six untrained musicians cannot pick up instruments and play jazz, in the same way that a battlegroup of army recruits cannot instantly understand a leadership concept as nuanced as Auftragstaktik.

You want to be the pebble in the pond that creates the ripple for change.

Tim Cook | CEO of Apple

NOW WHAT?

NOTHING ABOVE IS NEW, although it may be new to you. The next step, therefore, is to study and begin to educate yourself. Without laying out a detailed training plan, below are some things for you to consider in order to improve both your leadership and your team's performance.

Synchronization

Definitions are not enough. There is ample literature on this effect, and you should study it. Investigate how musicians, athletes and others learn to be "in the zone."

Trust

This bedrock element of leadership cannot be bestowed; it must be earned. Consciously lead by example, praising subordinates and allowing honest errors. Recall what I gained when I "took it on the chin" for my own subordinates. Empower your subordinates and encourage them to use initiative whenever possible. Remember the secretary's reaction to being empowered. Recognize more than just achievement. Praise effort, both privately and publicly, even when that effort falls short.

Leader's Intent

Clear expression of your intent is key. Encourage your subordinates to question you when you are unclear. Ask them to back brief you — not to gain permission, but to ensure that what they understand your intent. That their hearing matches what you *believe* you said.

Ensure that you always articulate the *why* of your mission and explain the end state, what success will look like. Along with *why*, avoid if at all possible, restricting the *how* of any task. This will expand your subordinates' freedom of action.

You are responsible to ensure that you allocate sufficient resources to accomplish the tasks assigned. When setting boundaries, be sure to ask if your subordinates feel they have enough resources as well as understanding their boundaries and restrictions.

Initiative Training

Schedule short training sessions where you present a problem and then urge your team members to offer solutions. Never dismiss a solution. Support critical reviews of the consequences and let others add their own ideas. In these sessions, the outcomes are always less important than the process of showing them that you trust them to show initiative. Put yourself in a subordinate role so someone who works for you can experience what initiative and leadership feels like.

Always schedule After Action Reviews to consider what you wanted to achieve; what was actually achieved; what went well; and what could be improved. Be sure to begin with your own shortcomings; how you could have done better to ensure success.

Learning to Let Go

Create small, decentralized teams and encourage them to self-organize to complete a task, while you step away and let it happen. Encourage them to crosstalk, to communicate without going through you. Consider the mistake I made as a young troop leader. The only thing worse than micromanagement is the "delegate and disappear" technique. Observe so that you can assist but do not intrude except to stave off failure.

Final Note

Keep in mind the old saw about how you get to Carnegie Hall: practice, practice, practice! Leadership, like music, is a blend of art and science that requires practice, making mistakes, and improvisation. Nobody becomes a world class jazz musician overnight and your team won't fully understand the concept of *tactical jazz* unless you embrace the challenges of trust, and offer subordinates more freedom so that together you learn how to build synergy.

I'm not a theoretician about playwriting, but I have a strong sense that plays have to be pitched - the scene, the line, the word - at the exact point where the audience has just the right amount of information. It's like Occam's razor.

Tom Stoppard | English Playwright

SHAVING WITH OCCAM'S RAZOR

MORE IS NOT NECESSARILY BETTER. Simple is superior to complex. The paradox that "less is more" is another way of describing the philosophy of Brother William of Occam, the fourteenth century English friar and his famous intellectual razor, which is also related to Zen Buddhism and Taoism. Occam's Razor and the Japanese concept of Zen share a common emphasis on simplicity, clarity, and the elimination of unnecessary complexity. In other words, cut away the unnecessary. A passage from my book, *Praxis Tacticum*, illustrates the concept.

Miyamoto looked at his student. "You have studied well; you are almost a novice." The student frowned. After so much hard work how could his beloved sensei say that he was a novice? The old sensei saw the disappointment. "You are troubled?" he asked.

"Master, I have studied with you since childhood and now you tell me that after all of my hard work, I am but a novice."

The sensei bade his student to sit. "Zen teaches us that thought, and action are one. When you first arrived, you acted instinctively. You acted as you thought but you were untutored in the ways of the hand, the pen, or the sword. As you learned the many techniques, thought and action grew distant from each other. You were clumsy and slow. You had to ponder momentarily each time you used the hand or the pen or the sword. Now that hesitation, the gap, has been removed; you once more act without hesitation. You behave instinctively. You behave as the novice, but with the sense of Zen, in the state of zanshin."

––––––––––

Too often, leaders tend to complicate what is simple. Sometimes this unnecessary embellishment stems from arrogance, but often it is done out of a misguided belief that complexity demonstrates knowledge. The opposite is true.

Above, the *sensei* praised his student by telling him that he had achieved a state of *zanshin*, or relaxed alertness, literally "the mind with no remainder." In other words, the student had achieved a state of complete and natural focus and a constant awareness of his body, mind, and surroundings. He had achieved the essence of effortless vigilance.

When we turn our minds to leadership, we know that we must create and then maintain an atmosphere of trust in order for our subordinates to thrive. Often, in an effort to do so, we complicate their lives, thereby weakening our organizations. This might be with complex mission statements, or through command and control structures that have unnecessary redundancy. Another way to state this is that we create systems with *too much information*.

As with *Auftragstaktik*, there is no quick or easy way to understand this philosophy. It takes study. But now, you have a point of departure and some suggestions on where to begin. To quote Saint Augustine, now it is time to "Take up and read!"

Zen Aesthetic

The Zen influence is evident in aesthetics which find truth in simplicity and beauty in imperfection. This thinking aligns with Occam's Razor in the sense that the simplest, most unadorned forms can reveal profound truths or beauty. Both Occam's Razor and Zen advocate simplicity and clarity, albeit in different contexts — Occam's Razor in its intellectual inquiry and Zen in its spiritual practice and daily living. Their shared principles reflect a universal appreciation of the elegance of simplicity.

Now this is not the end. It is not even the beginning of the end. But it is, perhaps, the end of the beginning.

Winston Churchill

ONLY THE BEGINNING

WE HAVE NOT REACHED A CONCLUSION. We have arrived at a beginning.

I've introduced some ideas on how to become a better leader so that you may achieve better cohesiveness, stronger relationships and enhanced outcomes for your organization. The fact that my examples are initially drawn from my military experience should not put you off. They'll work in any setting. I do not promise that this will be quick or simple. Remember what I said at the outset: I am but offering an *introduction* to an idea.

Consider all of these ideas to be like pebbles cast into a pond. The ripples may reach the shore, or they may fade amid the water. Perhaps two or three ripples will combine to form a small wave of inspiration. That is my hope for you.

Tolle Lege, tolle lege!
(Take up and read!)

St Augustine

FURTHER READING

HERE ARE SOME SUGGESTIONS TO CONSIDER. Note that they are not "how to" books:

Alan C. Greenberg. *Memos from the Chairman*, Workman Publishing, 1996

This book may not be for everyone. The Chairman of Bear Stearns has a simple message, and he tends toward repetition of that message. The memos are both entertaining and humorous, but the book has a limited applicability to modern business environments. Worthwhile, nonetheless.

Benjamin Hoff. *The Tao of Pooh*, Penguin Books, 1983

While criticized for its limitations, it remains a popular and influential book which has inspired many to explore Taoist philosophy and adopt more mindful and simple approaches to life.

—— *The Te of Piglet*, Penguin Books, 1993

Similarly to his previous work Hoff's simplified exploration of Taoist philosophy is not meant as a deep dive, but a thought-provoking and insightful exploration.

Miyamoto Musashi. *The Book of Five Rings*, Thomas Cleary ed. and trans., Boston: Shambhala, 1993

This seventeenth century classic guide to strategy has been widely read and applied in various fields, including martial arts, business, and philosophy. It is not a straightforward read, and on the surface some aspects may seem to be contradictory. It's real value is in its insights and principles.

Robert M. Pirsig. *Zen and the Art of Motorcycle Maintenance*, Mariner Books, 2008

Pirsig's complex and multifaceted nature challenges traditional philosophical assumptions and spark debate, instead of providing a comprehensive or systematic treatment of his concepts. The anecdotal storytelling and emphasis on individual experience helps to make complex philosophical ideas more accessible. Not many books are life-altering. For me, this one was.

Charles S. Oliviero, "Trust, Manoeuvre Warfare, Mission Command and Canada's Army," *The Army Doctrine and Training Bulletin* No. 1, Vol 1, August 1998

I was asked to write a guest editorial for the inaugural edition of what later became the Canadian Army Journal. In this short essay, I decry the attempt to graft a foreign concept onto Canadian doctrine without an in-depth analysis of what is required to adopt such a foreign idea — however useful and valuable the idea may be.

Acknowledgements

I would like to thank my old friend and high school mate, Angelo Mattachione for prompting me to pen this book, as well as my friend and publisher Phil Halton for pushing me to write it. It is good to have friends.

I would also like to thank my group of so-called Beta Readers for their time and their valuable comments and insights as well as those who took the time to offer their opinions in print.

THE COHORT

CHARLES S. OLIVIERO

The Cohort: Trust and Betrayal

Though slated for high command, Colonel Amadeus Ignazio "Skip" Schiaparelli abandoned his life in uniform when the Army no longer felt like home. Retiring to a contemplative existence in Italy, he is recruited by an American intelligence officer with a stirring proposal. Intrigued, if somewhat skeptical, Skip accepts the offer and is thereafter launched into a shadowy world of lies and deception which threaten many of his core beliefs.

Skip is a man committed to the ancient principles imparted to him by his father: trust, friendship and honour. He surrounds himself with friends, both old and new, who share those same beliefs. But even in Eden there was a snake, and betrayal threatens this new life of brotherhood and fraternal fidelity that he creates.

Whatever the challenge, Skip follows his guiding principle, that trust is earned not given. But in a world of deception, can one man trust his own instincts?

Auftragstaktik

✠ The Birth of Enlightened Leadership ✠

The Author of Praxis Tacticum and Strategia

Colonel Charles S. Oliviero

Auftragstaktik: The Birth of Enlightened Leadership

Auftragstaktik (mission-type tactics) were the foundation of Prussia's, and later Germany's, astounding battlefield performances. Universally praised in military circles, Auftragstaktik remains both poorly understood and badly practised. It is sometimes mistaken to be a military leadership strategy, when in fact it is a philosophy that applies to the exercise of leadership in nearly any situation. For many reasons, this uniquely German approach to leadership lies shrouded in a fog of mystery and misinterpretation.

Colonel Oliviero clears away that fog.

He concisely explains the early intellectual and structural growth of this leadership philosophy from before the French Revolution to the eve of the First World War, with a glimpse at the interwar period. He provides the necessary background and understanding to any reader, military or civilian, looking to learn about using mission-type tactics through the lens of history.

STRATEGIA

A PRIMER ON THEORY AND STRATEGY
FOR STUDENTS OF WAR

COLONEL CHARLES S. OLIVIERO

Strategia: A Primer on Theory and Strategy for Students of War

War fascinates us, but what do we really know about its nature?

Strategia (The Science and Art of Military Command) is the product of Colonel Oliviero's decades-long intellectual quest to address this fundamental query. His work offers both the serious student and the casual reader a foundation stone upon which to build a deeper understanding of military thought and theory, and thereby a richer appreciation of mankind's deadliest pursuit.

Strategia introduces many of the major contributors to military thought and theory as well as some of their most profound impacts on the conduct of war, from Sun-Tzu to the modern day, encompassing warfare on land, at sea and in the air, as well as in the cyber-realm.

While not an all-encompassing deep dive, Strategia is an essential primer and a point of departure. With this foundation stone in place, the student of war can proceed to follow Clausewitz's admonition to develop a "fine and penetrating mind."

PRAXIS TACTICUM

THE ART, SCIENCE AND PRACTICE OF MILITARY TACTICS

COLONEL CHARLES S. OLIVIERO

Praxis Tacticum: The Art, Science and Practice of Military Tactics

Pundits have long predicted the end of conventional warfare but for the foreseeable future, it is here to stay. Counterinsurgency, guerrilla warfare, terrorism, peace enforcement, policing.

All these forms, like conventional warfare, are as old as mankind. Modern militaries claim to be professional bodies, responsible for the education, control and discipline of their members. But at least one aspect of this claim is poorly executed: tactics are not taught to junior leaders, which is why this guide is essential reading for all junior leaders, officers and NCOs alike.

There is a military adage that there is no teacher like the enemy. True; but the wise commander prepares to meet that enemy and become their teacher instead. This is your essential study guide.

Notes

About the Author

Colonel Chuck Oliviero, PhD spent almost four decades in the Canadian Army as an Armoured Cavalry officer. He spent half of his career either as a student or as a teacher and the other half commanding troops. He is the author of three non-fiction books on war and strategy: *Praxis Tacticum*, *Strategia* and *Auftragstaktik*, as well as one novel, *The Cohort*. All are published by Double Dagger Books and available worldwide.

DOUBLE‡DAGGER

— www.doubledagger.ca —

DOUBLE DAGGER BOOKS is Canada's only military-focused publisher. Conflict and warfare have shaped human history since before we began to record it. The earliest stories that we know of, passed on as oral tradition, speak of war, and more importantly, the essential elements of the human condition that are revealed under its pressure.

We are dedicated to publishing material that, while rooted in conflict, transcend the idea of "war" as merely a genre. Fiction, non-fiction, and stuff that defies categorization, we want to read it all.

Because if you want peace, study war.

www.ingramcontent.com/pod-product-compliance
Lightning Source LLC
Chambersburg PA
CBHW071435210326
41597CB00020B/3796